· NORTHWEST ·
KNOW-HOW

TREES

NORTHWEST

KNOW-HOW

TREES

Karen Gaudette Brewer

Illustrations by **Emily Poole**

SASQUATCH BOOKS

SEATTLE

For Dad, who made sure we went camping, and for Mom,
who made sure we noticed everything around us
—KGB

Copyright © 2021 by Sasquatch Books

Printed in China

SASQUATCH BOOKS with colophon is a registered trademark
of Penguin Random House LLC

27 26 25 24 23   10 9 8 7 6 5 4 3

Text: Karen Gaudette Brewer
Illustrations: Emily Poole
Editor: Daniel Germain
Production editor: Bridget Sweet
Designer: Tony Ong

Library of Congress Cataloging-in-Publication Data
Names: Brewer, Karen Gaudette, author.
Title: Northwest know-how: trees / by Karen Gaudette Brewer, illustrated by
    Emily Poole.
Other titles: Northwest know-how.
Description: Seattle, WA : Sasquatch Books, [2021] | Series: Northwest
    know-how | Includes bibliographical references and index.
Identifiers: LCCN 2020037286 | ISBN 9781632173522 (hardcover)
Subjects: LCSH: Trees–Northwest, Pacific–Identification.
Classification: LCC QK144 .B74 2021 | DDC 582.16–dc23
LC record available at https://lccn.loc.gov/2020037286

ISBN: 978-1-63217-352-2

Sasquatch Books
1325 Fourth Avenue, Suite 1025
Seattle, WA 98101

SasquatchBooks.com

CONTENTS

NORTHWEST WONDERS

ACKNOWLEDGMENTS

This book would not be possible without the support and encouragement of my family. Thank you to my husband, Jerry, for keeping our sons (and cat!) busy so that I could complete this project amid school closures and a global pandemic. Thank you to our boys, Miles and Austin, for collecting pine cones with me and going on endless nature walks under the guise of "hunting for Pokémon."

Much appreciation to Jennifer Worick and Kerry Colburn with the Business of Books for lasting lessons in publishing. Thank you to my editor, Daniel Germain at Sasquatch, for his upbeat ways, and to my fellow author and friend Nicole Tsong for keeping me focused and hopeful. To the University of Washington: Thank you for requiring humanities majors to take science coursework. I took forestry and botany classes that sparked a lasting desire to ID everything in my sight line.

And much love to Olalla Elementary School, whose evergreens I played beneath for many happy years. Unlike my big brother, I knew to steer clear of the pine cone fights.

—KGB

I owe a great deal to Tom, Whitey, and the iNaturalist community for aiding and abetting my arboreal education. My deepest thanks go to Kelly, Jordan, and my parents for their generous support and advice.

I dedicate the illustrations in this book to all those who protect and maintain our public lands so that we may continue to appreciate what our fellow species have to share with us.

—**EP**

INTRODUCTION

To live in the Pacific Northwest is to dwell in a kind of benevolent mystery.

You might spend an entire week here before it dawns on you, during a sunbreak, that a dormant volcano has been holding court behind the clouds all along. While hiking you could easily wander past chanterelles like those you'll later savor at dinner, not recognizing their fanlike forms amid the camouflage of a plush carpet of pine needles. On the water you might paddle in solitude—or yelp in surprise as a harbor seal pops up its whiskered face to see what you're up to.

But the greatest secret keepers are our trees: tall, confident, constant. They shield our homes from neighbors, shade streams so the salmon can spawn in peace, muffle our voices with their branches, and offer our famously introverted population a retreat from our rapidly growing cities. The lush woods that fringe our towns and cities with every imaginable shade of green offer a warm, welcoming embrace and endless opportunities for exploration.

There are epic trees, ponderous evergreens far too fat to hug, bathed in mist or sun, nourished by abundant snowmelt. There are otherworldly trees, larches whose golden glow gleams atop mountain crests each fall. Walk along rocky beaches and you'll glimpse madrones, their oddly smooth bark peeling in strips as they stand sentinel over waterways that once ferried explorers. That whoosh you hear during winter is the wind combing the needles of hemlocks and pines, spruces and firs. Better pull up your hood: the rain may have paused, but you'll be shocked when giant drops eventually descend from limbs high above to trickle down your neck.

In town trees serve as landmarks. In the country the timberline forces us to hike ever higher to finally, finally (finally!) ascend beyond the conifer crowns to be able to glimpse the payoff: dazzling mountain lakes, a sweeping view of the valley below, a peek at Mount Baker or Adams or Hood, or Mount Rainier or Saint Helens or Shasta shimmering in the distance.

Our native trees loom large in our imaginations because they've seen more than many of us will ever see in our lifetimes. And when our time is past, they'll remain to witness the future.

We hope this book helps you feel even more at home on your next hike, ski run, or picnic now that you'll be better able to recognize your neighbors. A few notes on our region's quirks before we begin:

- We alphabetized the trees by their Latin name so that similar species appear together. Consult the index to find a tree by its common name (e.g., Douglas fir).
- Trees are either deciduous (drop their leaves each fall) or evergreen. However, not all evergreens are conifers (trees with cones). And not all conifers are evergreens.
- Our "cedars" are actually cypresses. Our most famous "fir" is not a fir at all. Oregon's beloved "myrtle" is actually laurel. Buckle up, yew!

Pacific Silver Fir

Like its European cousin, *Abies alba*, Pacific silver fir is a familiar presence along ski runs and hiking trails. Its short branches readily shed snow, building up a deep, snow-free well around its base. That's a favorite spot for snowshoe hares to shelter.

Wondering how to tell common Northwest firs apart? Pacific silver fir needles are straight with a top row that leans forward like ski jumpers and the side needles spread to both sides evenly. Noble fir needles are bent near the base like hockey sticks. Grand fir needles are flat and alternate in short and long pairs along its twigs. The Douglas fir isn't a fir at all! But that's another story for later (see page 96).

LATIN NAME: *Abies amabilis*

SIZE: Around 150 feet at maturity, with a tall and lean pyramid shape; they can live as long as five hundred years

IDENTIFYING FEATURES: Lustrous evergreen needles with silvery undersides and notched tips; barrel-shaped cones that grow to around 5 inches long and eventually turn deep purple

GROW REGION: From southern Alaska to Northern California, typically at elevations of 1,000+ feet

FUN FACT: *Amabilis* translates as "loveable" (or "lovely"), a descriptor also reflected in its French name, *sapin gracieux* ("graceful fir")

Grand Fir

Grand fir smells like Christmas. Its aromatic properties were well known by local tribes, who used boughs as air fresheners and burned them as incense and to ward off illness.

It's also sturdy: some of the grand firs along Barlow Road on the south side of Oregon's Mount Hood still bear rope burns from where pioneers on the Oregon Trail lashed or "snubbed" their wagons to slow the steep descent into the fertile Willamette Valley.

LATIN NAME: *Abies grandis*

SIZE: Mature trees can reach 200 feet; width ranges from 1.5–3.5 feet, though old growth on the rainy coast can reach 6.5 feet

IDENTIFYING FEATURES: Evergreen needles are waxy, deep green on top and silver below, and grow flat across one linear plane; cones sit upright like candles near the treetop and mature to a dark purple or blueish gray

GROW REGION: From southern British Columbia to Northern California

FUN FACT: It's the preferred home of pileated woodpeckers; decay from Indian paint fungus (*Echinodontium tinctorium*) creates a perfect hollow chamber for their roosts

Subalpine Fir

Gaze at the timberline and you'll recognize subalpine fir by its spire-like tops and reedy upper branches laden with cones. You won't find any on the ground unless a squirrel gnawed one down: they disintegrate on the tree. Its stately posture sets it apart from neighbor mountain hemlock (see page 120), which will be flexing in the wind.

Subalpine fir is a territorial tree of the Yukon, just one of the conifer-rich regions with a tradition to steep the needles in boiling water for a tea rich in vitamin C.

LATIN NAME: *Abies lasiocarpa*

SIZE: Shorter than most firs, typically no taller than 80 feet

IDENTIFYING FEATURES: Evergreen needles have a white stripe on top and stripes below; cones are 3 inches long and deep indigo

GROW REGION: Widespread throughout the West near the timberline of the coolest and wettest forests

FUN FACT: They are shade tolerant and often in the shadow of taller, sun-loving species such as Douglas fir

Shasta Red Fir

Recent DNA analysis confirmed a hunch held by some botanists and foresters that these trees are a hybrid of noble fir (*Abies procera*; see page 9) and California red fir, prompting some to refer to Shasta red fir as *Abies procera* x *magnifica*. These substantial, handsome trees stand as a reminder that our understanding of nature continues to evolve with each year that passes.

LATIN NAME: *Abies magnifica* var. *shastensis*

SIZE: Around 130–200 feet tall and as wide as the length of a bathtub

IDENTIFYING FEATURES: Evergreen needles with white on top and bottom, about 1 inch long that curve up like a hockey stick; large woody cones about 6–9 inches long, and reddish-gray bark at maturity

GROW REGION: In the mountains of southwest Oregon and Northern California between 5,000 and 7,000 feet

FUN FACT: You'll find this tree, also known as silvertip, thriving around Oregon's Crater Lake National Park

Noble Fir

Noble firs got their name from Scottish botanist David Douglas in the early 1800s during his Northwest tree classification spree. Douglas called it *Abies nobilis* for its upstanding appearance, but that name was taken. Nobles were renamed *Abies procera*, a nod to their loftiness.

Unlike most true firs, nobles offer strong lumber: the timber industry dubbed nobles "larches" for a time to protect it from being branded soft or flimsy. Larch Mountain, the extinct volcano outside Portland, is named not for larches but for its nobles.

LATIN NAME: *Abies procera*

SIZE: Nobles are the tallest true fir and typically live up to four hundred years; they can top 200 feet at maturity and grow as wide as a dresser

IDENTIFYING FEATURES: Needles are blue green, sturdy, and appear perfectly groomed by an unseen blow dryer, curving outward uniformly around each twig; huge khaki cones top 8 inches in length

GROW REGION: Mostly along Washington and Oregon mountain passes

FUN FACT: With symmetrical branches and stately posture, nobles are the quintessential West Coast Christmas tree

Vine Maple

Walk the woods in autumn and you'll easily spot vine maple by the fiery reds and golds of its dainty leaves agleam against the green-and-brown backdrop of its conifer and western sword fern neighbors. It sprawls its stems any which way necessary to stretch into the light. When they venture into arid regions, they're cozied up to ponderosa pine (see page 87).

You'll also likely glimpse it around the neighborhood: it's among the most popular native shrubs for landscaping.

LATIN NAME: *Acer circinatum*

SIZE: A shrub in shade or a tree up to 30 feet in open habitat, with a trunk up to a foot wide

IDENTIFYING FEATURES: Leaves have seven to nine distinct lobes and are from 1.5–2.5 inches long; bark is smooth and reddish brown

GROW REGION: Southwestern British Columbia to Northern California, typically on the west side of the Cascades

FUN FACT: *Circinatum* refers to its rounded, regularly lobed leaves

Ghost Forests

On especially stormy nights they would reveal themselves at low tide: giant weathered stumps covered in barnacles and sea life, remnants of an ancient Sitka spruce forest that loomed along the central Oregon Coast some two thousand years before. The Neskowin Ghost Forest is one of more than two dozen haunting groves that fringe the Northwest's windblown beaches, sentinels to long-ago tsunami, volcanic eruptions, earthquakes, and other major

triggers of dramatic sea level change that abruptly plunged these thriving coastal woods into saltwater, killing the trees.

To the north near Ocean Shores in Copalis Beach (Grays Harbor County, Washington), you can paddle your kayak up the Copalis River to see dead, bleached cedars that have lingered since the year 1700 when the massive 9.0 Cascadia earthquake thundered through the Northwest, triggered a tsunami across the Pacific in Japan, and dropped the forest 6 feet, turning it into a saltwater marsh. Like its brethren along the coast, this ghost forest is a reminder that the Northwest's temperate nature belies explosive potential below the surface.

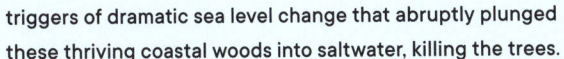

Douglas Maple

Acer glabrum is known as Rocky Mountain maple throughout the interior of North America and has several regional variations.

Of the three maple species native to the Pacific Northwest, only the big-leaf maple (see page 16) grows into a towering tree, while the Douglas maple and vine maple (see page 10) tend toward shrubbiness. Douglas maple's fall foliage rivals vine maple and ranges from crimson to orange to gold. When it's not busy brightening dark forests, it's an important source of food for deer, elk, and small mammals.

LATIN NAME: *Acer glabrum* var. *douglasii*

SIZE: Up to 30 feet tall as a tree; as a shrub, typically multistemmed

IDENTIFYING FEATURES: Smooth, toothed, deciduous leaves (*glabrum* means "without hair"), 1–3 inches wide, with three to five lobes

GROW REGION: Found from coastal southeast Alaska to southern Oregon, and east to Idaho and Montana

FUN FACT: Native groups used this tree's pliable branches for snowshoe frames, dip nets, and fishing hoops

Big-Leaf Maple

They weren't kidding about its leaves: they're a foot across, perfect for preschoolers to wave as makeshift flags. You'll find its wood in furniture and specialty guitars: after red alder (see page 19), big-leaf maples are the most abundant source of hardwood throughout the Northwest.

Big-leaf maples are declining, though tree pathologists aren't sure why. The stakes are high: with all their grandeur, they create habitat for a variety of plants and creatures, from nesting birds to licorice ferns.

LATIN NAME: *Acer macrophyllum*

SIZE: It's North America's biggest maple, around 100 feet tall when mature, with a trunk as wide as 5 feet

IDENTIFYING FEATURES: This deciduous tree offers plate-sized leaves that turn yellow come fall

GROW REGION: Along rivers and lakes throughout the Northwest; big-leaf maples adore moist soil and offer valuable shade to spawning salmon

FUN FACT: The sap of this West Coast native produces a dark amber maple syrup with notes of bourbon, caramel, and vanilla

Red Alder

Red alder are the table setters in Northwest forests, taking root in clearings or burned-over areas, eventually giving way to Douglas fir, western hemlock, and Sitka spruce (see pages 96, 119, and 67). While they elbow out competitors during their relatively brief lives, they give back wholeheartedly: red alder is a nitrogen fixer and its leaves, roots, and even its trunk (when it falls to the forest floor) all contribute nitrogen back to the topsoil that helps other growing things thrive.

Roasting and steaming salmon atop planks of alder and cedar are one of the traditional cooking methods of Northwest tribes to impart a subtle, woodsy nuance.

LATIN NAME: *Alnus rubra*

SIZE: They can top 35 feet in as little as a decade but live only forty to sixty years

IDENTIFYING FEATURES: Oval-shaped, serrated leaves that remain green come fall, with thin, speckled gray bark

GROW REGION: Along the Pacific coast from southeast Alaska to Southern California

FUN FACT: These fast-growing trees are a favorite construction material of beavers

Pacific Madrone

Next time you find yourself on a boat on saltwater, look for this lovely tree perched atop cliffs, its pearly white flowers and bright red berries standing out in a sea of green. But the peeling bark and smooth trunk is madrone's trademark, with colors ranging from orange to bright green.

The madrone is named for Archibald Menzies, a Scottish botanist whose name also is attached to the Northwest's most prolific tree, the Douglas fir (*Pseudotsuga menziesii*; see page 96). In Canada, it's called arbutus.

LATIN NAME: *Arbutus menziesii*

SIZE: When not stunted by coastal winds or poor soil these stately trees can grow 80–125 feet and several feet wide over 200–250 years

IDENTIFYING FEATURES: Evergreen, glossy, leathery ovoid leaves akin to coast rhododendrons

GROW REGION: From southwestern British Columbia through coastal California

FUN FACT: Madrone (or madrona) came from Spanish missionaries reminded of the similar red fruit of strawberry trees (*Arbutus unedo*) along the Mediterranean

Water Birch

If you find yourself unable to reach a stream to wade across or cast your fishing line, chances are you're being blocked by a thicket of water birch that's taken root along the shore. And thank goodness, as they reduce erosion and keep the waterways clear for fish and other creatures.

Their papery catkins (slim, nondescript, wind-pollinated flower clusters) feed grouse. Their inner bark feeds beavers. Their seeds feed chickadees and other small birds.

LATIN NAME: *Betula occidentalis*

SIZE: Around 30 feet tall and 1 foot wide; as a shrub, shorter than 20 feet, with multiple stems

IDENTIFYING FEATURES: Deciduous oval green leaves about 2 inches long, thin with serrated edges, turning gold come fall; bark is gray and matures to a red-brown hue

GROW REGION: In the Northwest, they congregate east of the Cascades near waterways, marshes, ravines, bogs, lake shores, and other saturated spots

FUN FACT: Water birch often hybridizes with paper birch (*Betula papyrifera*; see page 24)

Paper Birch

When Lewis and Clark set out in 1804 to learn as much as they could about all living things west of the Mississippi, then-President Thomas Jefferson advised them to record their observations on "paper of the birch" peeled from the trunks of trees he assumed they'd pass along the way.

Naturally waterproof and sturdy, it could preserve their journals through whatever weather might lay ahead. Unfortunately, their route traveled too far south to pass many stands, and the team was forced to record their learnings on "common paper."

LATIN NAME: *Betula papyrifera*

SIZE: This quick-growing tree reaches 70–90 feet during its 125 years, with a slender, straight trunk about 1 foot wide

IDENTIFYING FEATURES: Sharply pointed oval leaves with serrated edges that grow to about 2–4 inches long and turn buttercup yellow come fall

GROW REGION: The northern latitudes of North America

FUN FACT: Native tribes traditionally have used the bark to make canoes and baskets

A Cathedral of Giants

Most everyone is familiar with California's wondrous redwood groves, whose giant, ancient trees bathed in coastal mist continue to inch their crowns ever closer to the sun. But to the north, Vancouver Island is home to its own groves of colossal trees, where Douglas fir and western red cedar (see pages 96 and 116) that were saplings when Genghis Khan invaded China and King John signed the Magna Carta now filter the sun 250 feet above the ferns that unfurl along the forest floor.

The most famous of these woods is Cathedral Grove in MacMillan Provincial Park, which has been protected from logging by the Canadian government since 1947. But drive the (sometimes harrowing) perimeter of the island wedged between the Pacific Ocean and the Strait of Georgia and you'll get to commune with old-growth forest that also includes mature western hemlock, Pacific yew, red alder, big-leaf maple, and Pacific madrone (see pages 119, 113, 19, 16, and 20).

Worthy stops include Goldstream Provincial Park, just outside Victoria; Avatar Grove outside Port Renfrew, where you can also visit 20-story-high Big Lonely Doug, the last tree standing amid acres of stumps that remain from a 2011 clear-cut; and Carmanah Walbran Provincial Park, home to ancient spruce trees.

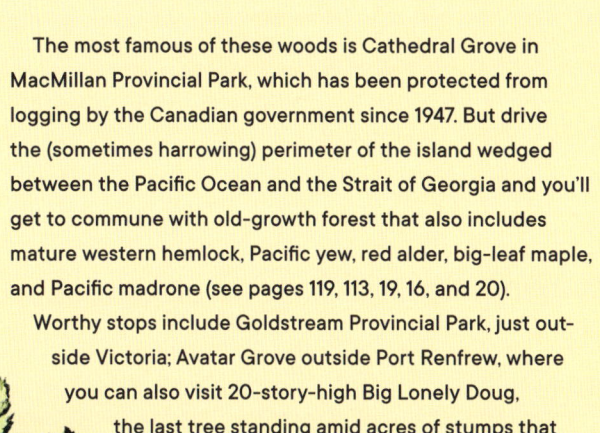

Incense Cedar

Before logging became widespread, incense cedars reportedly reached 225 feet tall, 12 feet wide, and lived over five hundred years. True to its name, this species offers aromatic wood and that eternal soothing presence that cedars—even cedars in name only—provide.

Is that perky, pyramid conifer over there an incense cedar or a young sequoia? Incense cedars have hard, cinnamon-red bark broken into irregular ridges, while sequoias have spongy red bark that shreds easily. Incense cedars have flattened scalelike needles; sequoias have more rounded, prickly needles.

LATIN NAME: *Calocedrus decurrens*

SIZE: This medium conifer grows 80–120 feet tall and up to 4 feet wide

IDENTIFYING FEATURES: Evergreen needles arranged in flattened sprays

GROW REGION: The mountains of Western Oregon south through the Sierra Nevada

FUN FACT: Like western red cedar (see page 116), incense cedar is actually a cypress; its fine-grained, warp-proof wood makes it ideal for making pencils

Netleaf Hackberry

Netleaf hackberry is sometimes the only treelike plant around in the arid habitats where it thrives, making it an essential source of nourishment for all manner of wildlife.

Robins, jays, squirrels, coyotes, and many other animals devour its fruit. Insects enjoy its leaves. Deer munch its young foliage. And creatures of all sizes seek its shade on brutally hot days.

LATIN NAME: *Celtis reticulata*

SIZE: These loner, scrubby, dryland trees rarely top 30 feet; they commonly live up to two hundred years

IDENTIFYING FEATURES: Slightly heart-shaped green leaves with raised veins on the underside and serrated edges; cherrylike, red fruit; gray, warty, rough bark that matures to a reddish gray

GROW REGION: In Washington and Oregon you'll typically find them only in the driest regions such as the Columbia and Snake River Valleys

FUN FACT: This resilient species is actually a member of the elm family

Mountain Mahogany

This hardwood fueled the steady, intense fires needed to smelt silver ore during the era of the Comstock Lode in the 1850s. Local tribes used it to make arrowheads and other tools and brewed tea from the tree's bark to treat colds and make a rose-colored dye.

Spring brings small whitish-yellow flowers. Its trademark fruit is a hard seed topped by a feathery plume that helps it sail along winds to disperse.

LATIN NAME: *Cercocarpus ledifolius*

SIZE: Up to 15 feet as a shrub or 50 feet as a small, slow-growing tree, with trunks that resemble the twisty-yet-graceful array of acacia trees on a Kenyan safari

IDENTIFYING FEATURES: Elliptical evergreen leaves up to 1 inch long in clusters, thick and leathery, with edges curled under

GROW REGION: Dry sites in southern Oregon

FUN FACT: Mahogany is a misnomer: it's a member of the rose family; its common name derives from its dense, heavy wood

Port Orford Cedar

Port Orford cedar—surprise!—is actually a sturdy cypress once used extensively for shipbuilding. Its lumber now is exported almost exclusively to Japan for use in home and temple construction due to its similarity to hinoki, a native Japanese tree.

A root fungus is slowly killing off this species in its native habitat, though it thrives in yards and gardens around the globe where it's been celebrated as an attractive cultivar since the 1800s.

LATIN NAME: *Chamaecyparis lawsoniana*

SIZE: Mature trees grow up to 200 feet tall with trunks as wide as a loveseat, and can live for more than six hundred years

IDENTIFYING FEATURES: Small, flattened, scalelike needles arranged close to their branchlets; pendulous branches give it a sense of graceful motion

GROW REGION: Between Oregon's Coos Bay to the Northern California coast, up to 5,000 feet

FUN FACT: Horticulturists insist this tree should be called Lawson cypress, while foresters prefer cedar

Golden Chinquapin

Chrysolepis means "golden scales." *Chrysophylla* means "golden leaves." In an earlier era, golden chinquapin was known as *Castanopsis chrysophylla*, or chestnut-like tree with golden leaves. That's a whole lotta gold, and it's spectacular in bloom.

Spring brings creamy white flowers in upright catkins that become chestnut-like spiny golden burrs in fall, each holding one to three nuts that are beloved by birds and squirrels.

LATIN NAME: *Chrysolepis chrysophylla*

SIZE: They slowly grow into narrow pyramids 30–100 feet tall

IDENTIFYING FEATURES: Glossy and dense lance-shaped evergreen leaves with fuzzy golden scales beneath

GROW REGION: Along Washington's Columbia River Gorge and Hood Canal through Northern California

FUN FACT: These trees with their golden fruit are the only known host in Washington State to the golden or chinquapin hairstreak butterfly, a candidate for protected status due to its dwindling food sources

Sculpted by Wind

They bow and bend but do not break: we're talking pines, firs, spruces, and other Pacific Northwest trees that endure frequent blasts of wind, ocean spray, and snow that have adopted a crooked, stooped stance known as krummholz (yes, the German language truly does have a word handy for everything).

Some call the krummholz effect the original bonsai, referring to the Japanese art of pruning trees and shrubs to adopt specific aesthetically pleasing shape. Indeed, encountering the

wizened forms of mountain hemlock, western larch, subalpine fir, whitebark pine (see pages 120, 56, 5, and 70), and other alpine favorites on a foggy morning brings to mind fairy tales and fantasy novels where an elf or ogre may be just around the bend.

Don't be fooled by the small stature of some especially gnarled krummholz trees: though shorter, they may be considerably older and hardier than their taller brethren who've had the good fortune to grow freely toward the sun.

Pacific Dogwood

Pacific dogwoods are the wild, taller cousins of the dainty dogwoods that dwell in many urban yards, with striking clusters of white bracts that bloom in spring and often again later in the year. Come fall, their leaves turn a vibrant pink. You'll find them cozied up to streams and along gentle slopes.

You might already possess items carved and crafted from its hard, heavy wood at home: mallet handles, piano keys, cabinets, thread spindles, and bows. Its blooms grace the flag of British Columbia, Canada's westernmost province.

LATIN NAME: *Cornus nuttallii*

SIZE: Around 30–60 feet tall

IDENTIFYING FEATURES: Deep-green oval leaves with a pointed tip and parallel veins, with berrylike reddish-orange fruit in autumn

GROW REGION: Throughout the Pacific Northwest, primarily west of the Cascades

FUN FACT: This species gets its Latin name from Thomas Nuttall, an English botanist and ornithologist who shared his observations with John James Audubon

Black Hawthorn

Of the hundreds of hawthorn species in North America, Europe, and Asia, black hawthorn (also known as Douglas hawthorn) is the most ubiquitous of its genus. The widespread roots of these hardy trees help prevent soil erosion; the thickets created when the species takes on a more shrublike shape provide excellent cover and nesting sites for wildlife.

Spring brings apple-like white blossoms that are globe-shaped in small clusters. Its reddish-purple fruit is beloved by birds in the chillier months.

LATIN NAME: *Crataegus douglasii*

SIZE: Up to 35 feet as a small tree and shorter as a large shrub

IDENTIFYING FEATURES: Deciduous leaves are a shiny dark green, long, fan-shaped, and serrated at the tip; the tree has sharp, slender thorns and shaggy bark

GROW REGION: From southeastern Alaska to Northern California, typically at lower elevations

FUN FACT: Indigenous groups used the sturdy thorns as fishing hooks and sewing needles

Modoc Cypress

The Latin name of this species honors Milo Samuel Baker, the California botanist who classified this species in 1898.

And, like other western members of the cypress family, its Latin name is in dispute within the forestry and botany communities. It's also classified as *Hesperocyparis bakeri*, to reflect that it's a western cypress that grows in an incredibly specific section of the West.

LATIN NAME: *Cupressus bakeri*

SIZE: This slow-growing cypress eventually reaches 50 feet tall, though if you planted it, it wouldn't reach that height in your lifetime

IDENTIFYING FEATURES: Evergreen, scalelike leaves with circular glandular dots and white resin, with aromatic twigs

GROW REGION: The very southern extent of Oregon's Siskiyou Mountains, in rocky dry sites at higher elevation with poor soil

FUN FACT: Its small, round cones only open and release their seeds after their parent tree is killed by wildfire

Yellow Cedar

This lovely life-giving tree knows well who it is, but its admirers have given it an identity crisis. Like its fellow Northwest "cedars," it's actually a member of the cypress family. Depending where you live, it's known as everything from Nootka cedar to Sitka cypress to Nootka false cypress. It has borne four separate Latin names since Scottish botanist David Don first spied one near Vancouver Island's Nootka Sound.

First nations have long recognized its value for crafting everything from robes to masks to paddles to dishes. But for the past century, yellow cedar has been dying off in large stands. One culprit: less snow and earlier snowmelt each spring, which has left shallow roots uninsulated against arctic winters and susceptible to damage. As with glaciers, the time to savor this iconic tree is now.

LATIN NAME: *Cupressus nootkatensis*

SIZE: Around 100–125 feet tall; the eldest live more than a thousand years

IDENTIFYING FEATURES: Flattened sprays of needles on droopy branches; bark is grayish brown and pulls off in flakes

GROW REGION: At sea level from Alaska to Oregon, west of the Cascades

FUN FACT: You can tell it apart from western red cedar (see page 116) by stroking a branch away from its tip (prickly means yellow; smooth means red)

Oregon Ash

Fraxinus latifolia is the Pacific Northwest's only native ash and can often be seen in swampier parts, creating magical reflections when its leaves shine golden in the fall.

This hardwood provided strong timber for canoe paddles and digging implements for native tribes. Oregon ash is now used to make tool handles, furniture, sports equipment, and barrels.

LATIN NAME: *Fraxinus latifolia*

SIZE: Oregon ash grows quickly for its first century, reaching 60–80 feet, then continues on slowly for another hundred centuries if the conditions are right

IDENTIFYING FEATURES: Light-green deciduous compound leaves with five to seven leaflets; its inconspicuous flowers appear before its leaves; seeds are large, drooping clusters and resemble half a maple seed

GROW REGION: From the southern coast of British Columbia south to California, most common in valleys and along rivers

FUN FACT: The Latin *latifolia* means "wide leaves"; Oregon ash has wider leaves than most ash species

Western Juniper

Forget about the low-lying ornamental junipers you've seen in gardens, smelling faintly of gin: western junipers are their superhero brethren, growing tree-sized in the arid climates of central Oregon and Eastern Washington.

Many you see today have been growing steadily since the late 1800s and spreading rapidly in an era of fire suppression, making ranchers worry these plants will further eliminate grazing lands. But all manner of wildlife appreciates juniper berries for forage come winter.

LATIN NAME: *Juniperus occidentalis*

SIZE: They reach 30 feet after eighty to a hundred years, but can seem even larger as they often go it alone

IDENTIFYING FEATURES: Outstretched evergreen scales with white glands beneath; bark is reddish brown and stringy; small blueberry-like cones

GROW REGION: The sections of Washington and Oregon where it barely rains (yes, they exist)

FUN FACT: All junipers are members of the cypress family (Cupressaceae), which in the Northwest also includes western red cedar, incense cedar, Port Orford cedar, coast redwood, and yellow cedar (see pages 116, 28, 35, 110, and 47)

Alpine Larch

Larch lumber has little commercial value. But hikers and nature photographers treasure larches and trek for miles for the opportunity to glimpse and admire their strikingly golden needles each fall.

The biological clock of the alpine larch (a.k.a. subalpine larch) does not seem to be ticking loudly; these larches, which typically live four to five hundred years, don't produce cones until they are over one hundred and don't produce seed in any quantity until they are two hundred years old.

LATIN NAME: *Larix lyallii*

SIZE: Around 40–50 feet high and up to 2 feet wide

IDENTIFYING FEATURES: Needles are grouped into clusters of thirty to forty; cones are reddish yellow to purple, 1.5–2 inches long

GROW REGION: High in the Cascades and Rockies

FUN FACT: Fellow high-elevation creatures rely on larches: blue grouse consume its needles; mountain goats, pikas, and snowshoe hares rely on them for summer habitat; and woodpeckers make homes in larger specimens

Larch Madness

New Englanders famously go leaf peeping come fall, admiring the brilliant reds, oranges, and yellows of sugar maples and other hardwoods. But each October in the Northwest, it's time for larch madness, when scores of hikers, photographers, and tree lovers navigate mountain passes and steep terrain to witness these unusual deciduous conifers, their needles a brilliant gold, gleaming against a backdrop of evergreens, alpine lakes, and sometimes even snowy peaks.

The otherworldly display draws tourists from around the region, akin to the groups who pilgrimage to the Rockies each fall to witness the autumn gold of quaking aspen before they drop their tremulous leaves for winter.

In Washington, the North Cascades Highway is your best bet, though likely also the most crowded option. Call ahead to the ranger station or the visitor center during larch season for the latest information on peak foliage. In Oregon, check out the Elkhorn Scenic Byway along I–84 east of Pendleton. Check with your local hiking group for more larch peeping tips.

Western Larch

Dropping their needles gives western larches two key survival advantages: their ability to recycle nutrients like nitrogen helps them thrive in nutrient-poor soil, and lack of needles helps them better manage heavy snow loads that could otherwise snap bare branches.

No wonder they can live so long—up to eight hundred years.

LATIN NAME: *Larix occidentalis*

SIZE: The world's tallest larch, up to 150 feet with a trunk 3 feet wide; they would tower over alpine larch (see page 52) further up the mountain

IDENTIFYING FEATURES: Spring-green needles grow in bunches and drop off each fall after turning gold; abundant petite oval cones with bristly, spiky bracts

GROW REGION: The interior Pacific Northwest, often near Douglas fir, ponderosa pine, paper birch, quaking aspen, and black cottonwood (see pages 96, 87, 24, 88, and 91)

FUN FACT: Larches are conifers (their seeds grow in cones) but not evergreens; their leaves turn golden each fall and drop off their branches, typically beginning around mid-October

Pacific Crab Apple

Beyond providing wood for harpoons, fruit for hungry bellies, and flowers to stand out against its green conifer neighbors, this tree also provides habitat for as many as four-dozen species of butterflies and moths and helps to stabilize soils along stream banks throughout the Northwest.

LATIN NAME: *Malus fusca*

SIZE: It can reach up to 30 feet but typically hovers around 10–15 feet and is generally wider than tall

IDENTIFYING FEATURES: Deciduous leaves with a lobe on either side, deep green above, paler below; delicate white blossoms in spring give way to oblong, sour apples in the fall that are small and yellow to purplish red; autumn leaf hues range from reddish to orange

GROW REGION: West of the Cascades crest from Alaska to California in moist woods, swamps, and open canyons

FUN FACT: This is the only apple tree endemic to the West Coast and is also known as Oregon crab apple

Tanoak

Is it an oak? A beech? An evergreen? Tanoaks contain multitudes. They're an evergreen hardwood in the beech family with characteristics of oak and chestnut. Their closest relatives hail from southeast Asia. They're tenacious, yet susceptible to sudden oak death, a disease that is ravaging their population.

They have been alternately valued for their acorns, their lumber, and for the tannins contained in their bark that were used to treat leather. They've been treated as weeds to clear space for commercially valuable conifers; now, foresters and ecologists are racing to preserve remaining tanoaks for their beauty and enduring appeal.

LATIN NAME: *Notholithocarpus densiflorus*

SIZE: Around 50–150 feet tall and as wide as a fridge; they can live as long as 250 years

IDENTIFYING FEATURES: Thick and leathery evergreen leaves 3–5 inches long with bluish-white fuzz beneath; bark is thin with flattened ridges or plates; seeds resemble oak acorns, with spiked caps

GROW REGION: Southwestern Oregon south into California

FUN FACT: In its range, tanoak is second to Pacific madrone as the most abundant hardwood

Brewer's Weeping Spruce

As a child, these slow-growing evergreens with their pendulous, swoopy, "weeping branches" struck me as mischievous Dr. Seuss characters prepared to scamper off into the woods the instant we turned our backs. In reality, I'm lucky to have spotted one in the wild at all: they are among the rarest of American spruces.

You're more likely to spot one weeping in a neighbor's yard, flailing its evergreen "arms" on breezy days.

LATIN NAME: *Picea breweriana*

SIZE: Up to 125 feet tall . . . eventually

IDENTIFYING FEATURES: Flattened evergreen needles up to 1 inch long along curtained branches swooping dramatically; cylindrical red-brown cones, 2.5–5 inches long

GROW REGION: Native to southwestern Oregon and Northern California

FUN FACT: Outside its native range, this soft-needled spruce is grown as a valuable ornamental for large spaces and is especially popular in the United Kingdom for parks and estates

Engelmann Spruce

This tall and handsome spruce does not hog the limelight or any other kind of light: you'll often find it growing under a canopy of large trees where it holds its own with other shade-tolerant conifers such as subalpine fir (see page 5).

Like other spruces, Engelmann spruce is used for lumber, making paper, and to create musical instruments including guitars and violins. Its common names also include silver spruce, white spruce, and mountain spruce.

LATIN NAME: *Picea engelmannii*

SIZE: From 45–130 feet with a regal, pyramid shape

IDENTIFYING FEATURES: Sharp, thin evergreen needles that are four-sided and grow around the twig like a bottle brush; cones are 3 inches long with paper-thin scales

GROW REGION: In the Cascades and eastern mountains around 3,000–6,000 feet; its cousin, the Sitka spruce (see page 67), grows at lower elevations

FUN FACT: This spruce is named after German-American botanist George Engelmann, who was one of the preeminent American naturalists of the nineteenth century

Sitka Spruce

Sitka spruce can seem impossibly wide when you first encounter them. A big reason is many seedlings are nourished by nurse logs on the forest floor, which supercharges growth for the next several hundred years. Old-growth giants reside in the temperate rain forests of Olympic National Park, where officials safeguarded them from being felled during World War II for the war effort.

Its strong, lightweight wood makes it ideal for shipbuilding, racing shells, pianos, and, significantly, airplanes. It has an ideal strength-to-weight ratio and ready availability of knot-free boards with a straight grain.

LATIN NAME: *Picea sitchensis*

SIZE: The world's tallest spruce, nourished by moist ocean air and summer fog, sometimes topping 200 feet

IDENTIFYING FEATURES: Sitka spruce needles are incredibly poky; cones are 3 inches and reddish to yellowish brown

GROW REGION: Coastal Alaska to Northern California

FUN FACT: Sitka spruce is the state tree of Alaska and named for Sitka Sound

A Forest You Must Dive to See

Between Seattle and Bellevue, submerged in the dark, chilly waters of Lake Washington, lies not one, not two, but three groups of old-growth trees—ancient reminders of the massive earthquakes that can rattle the region.

The *Seattle Times* reported that the one that shook Puget Sound around a thousand years ago triggered avalanches in the Olympic Mountains, shoved Bainbridge Island higher above sea level, sent a tsunami careening toward Whidbey Island, and dislodged entire forests, sending the trees sliding 200 to 400 yards into Lake Washington.

The trees are well preserved due to the low oxygen levels at the lake's bottom, many still standing upright after all these years as their roots and soil slid right along with them: in the 1990s, one logging company got charged with piracy for illegally salvaging the timber, which belongs to the state. Meanwhile, the trees have become part of the lake's habitat, along with crashed World War II–era airplanes, at least a dozen coal cars from a sunken barge, and scores of sunken boats. Good thing it's a big lake.

Whitebark Pine

Whitebark pine nuts have fed generations of Clark's nutcrackers and red squirrels, which gather caches of the nuts and pine cones, called middens. A single nutcracker can cache nearly one hundred thousand seeds in a good crop year; a squirrel's cache can contain more than one thousand nut-bearing cones.

Wondering how to tell this pine apart from look-alike limber pine (see page 77)? Smell the cones: the immature purple cones of whitebark give off a slightly sweet scent, while the immature green cones of limber pine smell of turpentine.

LATIN NAME: *Pinus albicaulis*

SIZE: Shrublike under challenging conditions and 40–60 feet tall and 2–5 feet across when thriving; can live up to five hundred years

IDENTIFYING FEATURES: Evergreen needles 1–3 inches long in bundles of five; woody cones 2–3 inches long, round, and purplish

GROW REGION: From 7,000–12,000 feet along West Coast mountain ranges

FUN FACT: Whitebarks can withstand hurricane-force winds on their mountain perches; you'll often find them gnarled in krummholz form (see page 38), determined to persist

Knobcone Pine

Unlike many conifers, these scrappy pines don't drop their lower branches as they mature. This enables forest fires to quickly reach their crowns and intensify to a temperature that melts the resin, scattering the forest floor with the next generation of knobcones.

The cones also have been known to become embedded within growing branches and trunks, enabling the seeds to remain fully viable sometimes decades later when the tree itself is finally destroyed by fire. The dense cones are a favorite of jewelry artisans who cure and burnish cross-sections of their interiors to a natural glow.

LATIN NAME: *Pinus attenuata*

SIZE: Typically 30 feet, though some can stretch to 100 feet

IDENTIFYING FEATURES: Needles in bundles of three, 3–7 inches long, slender and twisted; woody cones with knob-like bumps on one side

GROW REGION: The dry, rocky soil of the mountains of southwest Oregon

FUN FACT: *Attenuata* refers to this pine's slenderly tapering branches

Shore Pine

This resilient pine has many names: shore pine, lodgepole pine, twisted pine, contorta pine. Whether bent by prevailing winds along the coast or growing among firs and hemlocks in the Cascades, this twisty, skinny evergreen shrub is a survivor, and among the first to take root after forest fires.

One reason: the resin in their cones only melts when temperatures crest 113–140 degrees F, giving their seeds the best chance to literally rise from the ashes on the forest floor.

LATIN NAME: *Pinus contorta*

SIZE: Up to 70+ feet

IDENTIFYING FEATURES: Short yellowish-green or dark-green needles in bundles of two, with inch-long, egg-shaped cones

GROW REGION: Along the Pacific coast and amid Northwest mountain ranges

FUN FACT: Also known as lodgepole pine, as its straight and slender poles were perfect for supporting lodges

Limber Pine

Limber pine won't win many beauty contests, but it has earned the respect of miners, ranchers, and writers throughout the West for its tenacious spirit.

"Stop a moment and listen to the whistle of the mountain wind in its short, stubby, thick needles. . . . They are lifting their voices in a tale of their life that is all fierce endurance," wrote Donald Culross Peattie in his seminal book *A Natural History of Western Trees*. "A tree better adapted to endure timberline conditions does not exist."

LATIN NAME: *Pinus flexilis*

SIZE: Usually under 50 feet tall and 2 feet wide, often shrubby

IDENTIFYING FEATURES: Evergreen needles in bundles of five, 2–3 inches long with white lines; woody cones 3–7 inches long

GROW REGION: In the northeast corner of Oregon along dry, rocky ridges

FUN FACT: Limber pine's limber twigs help them manage heavy snowpack and high mountain winds and can be tied into knots

Jeffrey Pine

Wondering whether you're admiring a ponderosa or Jeffrey? Their cones hold clues. Ponderosa pine cones feature prickles that arch outward; Jeffrey pine cones prickles curve more inward. Watch your fingertips either way: ouch!

Like ponderosas (see page 87), Jeffrey pines offer a surprising fragrance described as reminiscent of pineapple, vanilla, lemon, or butterscotch. They're named for Scottish botanist (there were so many!) John Jeffrey, who followed in the footsteps of David Douglas to catalog the botanical bounty of the West.

LATIN NAME: *Pinus jeffreyi*

SIZE: It can reach hundreds of feet tall and 4–6 feet wide over several hundred years of growth

IDENTIFYING FEATURES: Needles grow in bundles of three, 6–10 inches long

GROW REGION: Jeffreys grow up to the timberline from southwest Oregon down through the Sierra Nevada

FUN FACT: Their hefty pine cones are prized for craft projects and their bark grows in brown plates that can resemble puzzle pieces

Sugar Pine

These stately pines get their name from the sugary resin that exudes from the freshly cut wood. Their lumber is valued for its stability, workability, and how quickly it grows.

Scottish botanist David Douglas, who marveled at many of the Northwest's native trees when he first encountered them in the 1800s, called sugar pine "the most princely of the genus."

LATIN NAME: *Pinus lambertiana*

SIZE: They're the biggest pine in the world, up to 200 feet; sugar pine would shop at the big and tall store much of its five hundred years on earth

IDENTIFYING FEATURES: Needles are 2–4 inches long in bundles of five

GROW REGION: From the mountains of southern Oregon to California, from 2,000–9,000 feet

FUN FACT: Mature cones are enormous (10–20 inches long!), treasured by collectors

The Lower 48's Very Own Rain Forests

Say "rain forest" and most thoughts turn to the Amazon and other biodiverse tropical jungles. But what makes a rain forest is not its heat, but its literal raininess. The lush forests in the Quinault, Queets, Hoh, and Bogachiel Valleys in Washington State's rainy northwest corner offer some of the best remaining examples of primeval temperate rain forest in the Lower 48.

You'll definitely want to pack your rain gear to visit Olympic National Park, where old-growth stands of Sitka spruce, western red cedar, big-leaf maple, Douglas fir (see pages 67, 116, 16, and 96), and other members of these complete forests hold court over quiet, moss-shrouded woodlands as driftwood logs pound the rocky beaches to the west. What nourishes this ecosystem?

About 12–14 feet of rain each year, moderate temperatures, and dead trees that take centuries to decay back to the soil, nourishing generations of flora and fauna around them all the while.

These rain forests once stretched from southern Oregon to southeast Alaska. Other temperate rain forests remain in isolated spots around the globe, including Chile, New Zealand, and Tasmania.

Western White Pine

These lovely, fluffy pines with their pyramid shape are the state tree of Idaho and a standout among Northwest pines for their large cones and distinctive bark. Its soft, pliable wood is often used to make matches or for wood carving.

It hates the shade and relies on forest fires and logging to help clear overshadowing neighbors out of the way so it can flourish.

LATIN NAME: *Pinus monticola*

SIZE: Up to 150 feet tall and 3 feet wide; they can live up to four hundred years

IDENTIFYING FEATURES: Evergreen needles in bundles of five, 2–4 inches long and soft; cones 6–10 inches long and curved like a banana; branches have a slightly upswept, buoyant appearance

GROW REGION: At middle elevations throughout the Cascades down to sea level in northwest Washington

FUN FACT: *Monticola* is Latin for "mountain dweller"

Ponderosa Pine

The more arid climate of the inland Northwest abounds with pines, but most stately and rugged among them is the ponderosa. It's a key lumber tree, second only to the Douglas fir, and used to frame buildings everywhere. It protects hillsides from erosion, shelters songbirds, and can withstand most forest fires. "Of all pines, this one gives forth the finest music to the winds," wrote famed naturalist John Muir.

You may want to snag one of their plump pine cones as a souvenir, but watch your fingertips lest its sharp prickles snag you first.

LATIN NAME: *Pinus ponderosa*

SIZE: From 60–100 feet tall with trunks 2–5 feet across at maturity

IDENTIFYING FEATURES: Needles grow in bunches of three and cones are 3–6 inches long

GROW REGION: Throughout the inland west, from British Columbia through the Sierra Nevada

FUN FACT: Like their Jeffrey pine relative (see page 78), sun-warmed ponderosa pines smell of vanilla, lemon, or butterscotch

Quaking Aspen

It's an unmistakable sound: an intense rustling akin to millions of tiny hands clapping for an unseen breeze. Fans travel from around the globe to bear witness before vibrant yellow leaves twinkle down into fields of gold each fall.

Aspens are aggressive pioneers and are quick to colonize burned areas. What it doesn't like is shade, which makes sense for a species whose own whitish bark carries out photosynthesis.

LATIN NAME: *Populus tremuloides*

SIZE: From 20–80 feet tall and 3–18 inches wide; they generally don't live more than 150 years as individuals, but can persist thousands of years as a collective root system

IDENTIFYING FEATURES: Leaves are bright green to yellowish green and roundish, and 1.5–3 inches wide with a pointed apex

GROW REGION: East of the Cascades at 5,000 feet or higher

FUN FACT: One theory behind why aspen "quake" (flutter their leaves on any breath of air) is to protect their leaves from insects

Black Cottonwood

Each spring the seeds of these prolific trees float by the millions like tiny clouds on the breeze and atop the water, an ephemeral migration of cotton balls akin at its peak to driving through a fluffy snowstorm.

LATIN NAME: *Populus trichocarpa*

SIZE: They're the largest broad-leaved tree in western North America and grow extremely fast when young, reaching heights of 200 feet

IDENTIFYING FEATURES: Glossy green deciduous leaves are pale on the bottom and range in shape from lance-like to triangular, turning a brilliant gold in the fall; bark is smooth and whitish when young and matures into gray with thick, deep furrows

GROW REGION: Cottonwoods adore moist environments; you'll even spot these growing in or alongside lakes; they grow in nearly every region of the Pacific Northwest

FUN FACT: Most cottonwood is used for pulp and paper due to its softness, light color, and ease of bleaching

Bitter Cherry

As its name suggests, the fruit of this tree is incredibly bitter and inedible to humans, though small mammals, rodents, various birds, and black bears flock to enjoy the bitter cherries each summer. They spread the pits far and wide, where they lie in wait for optimal conditions— sometimes decades—before germinating.

LATIN NAME: *Prunus emarginata*

SIZE: From 6–45 feet tall, depending on conditions during its short, quick-growing life; it typically fades as the canopy fills in above and its light sources dwindle

IDENTIFYING FEATURES: Petite deciduous leaves are finely toothed and rounded at the tip and turn yellow in fall; whitish-pink flowers in a flat-topped cluster produce fruit that is bright red and about a centimeter wide

GROW REGION: Up and down the Pacific coast and inland into Idaho, often along streams

FUN FACT: This nectar in the blossoms of this member of the rose family provides nourishment to butterflies

Klamath Plum

This tree was much valued when settlers first arrived in the region, recognizing, as tribes long had, that these plums offer important nourishment (what we now know as vitamins C and A, and fiber) whether fresh or dried as a garnish for game. The native trees were used as stock to graft other plum species they'd brought along to cultivate for future gardens.

LATIN NAME: *Prunus subcordata*

SIZE: Up to 25 feet as a small tree, but often seen as a shrubby thicket

IDENTIFYING FEATURES: Deciduous oval leaves are deep green, small, and faintly serrated; pink or white flowers in spring produce tart but edible Klamath plums, which are about 1 inch wide and ripe late in the season

GROW REGION: Southern Oregon and California, in forests at low elevation, but occasionally in the Cascades

FUN FACT: Klamath plum is a member of the genus *Prunus*, which is shared by cherries, peaches, and other stone fruit

Douglas Fir

Douglas firs may be the Northwest's most iconic tree, and its most misunderstood. It's been called a spruce, a pine, a hemlock, a fir. Turns out, it's its own whole thing, with brethren in other climates. The type that grows locally turns out to be the standard bearer for the whole genus. Its name reveals its variability: *Pseudotsuga* means "false hemlock." Its wood is valued for its strength and rustic beauty.

This species continues to hold the Guinness record for the world's tallest cut Christmas tree: in 1950 promoters of the then fledgling Northgate Mall painstakingly trucked a 221-foot Douglas fir from the Mount Rainier foothills to Seattle.

LATIN NAME: *Pseudotsuga menziesii*

SIZE: They regularly top 200 feet, with mammoth trunks

IDENTIFYING FEATURES: Soft evergreen needles with distinctive thick, furrowed bark

GROW REGION: From British Columbia to California, from sea level on up to 5,500 feet

FUN FACT: It's the tree on Oregon's state flag

Canyon Live Oak

Canyon live oak has a symbiotic relationship with fire. While a forest fire can wipe out individual trees, without fire, a stand of oaks may never achieve tree form, as fire triggers its roots to sprout.

The Latin *chrysolepis* means "golden scale" in reference to the yellowish acorn cap, or cup. Canyon live oak's acorns are egg-shaped and 0.5–2 inches long.

LATIN NAME: *Quercus chrysolepis*

SIZE: Up to 15 feet tall as a shrub or 80 feet in tree form, and up to 2 feet wide, depending on growing conditions

IDENTIFYING FEATURES: This broad-leaved evergreen grows two distinctive types of dark green, shiny leaves, some with smooth edges and others spiked like holly

GROW REGION: In the mountains and canyon bottoms of southern Oregon to California

FUN FACT: Leaves have a yellow, feltlike fuzz on the lower surface

Oregon White Oak

When settlers first rolled into the Northwest, they observed vast grasslands where oaks loomed sporadically as the dominant trees. Summertime forest fires, grazing, and managed burns by local tribes enabled these oaks with their thick, mostly fire-resistant bark to thrive. The Northwest has since lost about 95 percent of that prairie habitat.

You can get a sense of this unique ecosystem at Oregon Garden in the Willamette Valley, where a four-hundred-year-old Oregon white oak taller than the White House presides over a 25-acre native oak grove.

LATIN NAME: *Quercus garryana*

SIZE: Up to 80 feet tall and 3 feet wide

IDENTIFYING FEATURES: Deciduous shiny and bright green leaves with seven to nine rounded lobes and dainty acorns; fall color is a saddle brown

GROW REGION: Southwestern British Columbia through California

FUN FACT: Sturdy and attractive, Oregon white oak lumber is used in flooring, furniture, and wine barrels

California Black Oak

These iconic trees often stand alone on hills for two reasons: they dislike shade, and throughout history, local tribes burned competing weeds and saplings to ensure easy access to more abundant crops of acorns.

Their legacy serves as a reminder that mankind has shaped the landscape for far longer than many of us realize.

LATIN NAME: *Quercus kelloggii*

SIZE: The largest mountain oak in the West, reaching 50–100 feet tall; mature oaks (ninety+ years old) can top 100 feet

IDENTIFYING FEATURES: Bright green leaves with six lobes, 2.5–7 inches long, gold in the fall

GROW REGION: Southwestern Oregon south through California; it's the most abundant oak on the West Coast

FUN FACT: Its acorns are a traditional source of protein and continue to be harvested; they're also a favorite snack of bears, deer, woodpeckers, and squirrels

Cascara Buckthorn

Cascara buckthorn thrives in the understory along rivers and streams, its graceful branches offering a perch for birds and other wildlife beneath the rafters of the towering conifers in its orbit.

Yet it's most famed for the medicinal properties of its dried bark, which wowed Spanish missionaries and settlers alike when shared by local tribes. Cascara bark harvesting was and remains a cottage industry in many small towns throughout the West.

LATIN NAME: *Rhamnus purshiana*

SIZE: Around 20–30 feet as a small tree or up to 15 feet as a shrub with multiple trunks

IDENTIFYING FEATURES: Deciduous oblong leaves 2–6 inches long that turn yellow to orangish-red come fall; fruit is purplish black

GROW REGION: From British Columbia to Northern California, typically west of the Cascades

FUN FACT: Its bitter bark has brought relief to millions in its traditional use as a laxative; small wonder one of its nicknames is "sacred bark"

Pacific Willow

You'll likely encounter this fast-growing tree on your next hike or fishing trip, legion as it is along the region's rivers. Beavers use it to build dams and lodges. If salmon notice such things, they'd appreciate its ability to prevent erosion along streambeds.

Due to extensive wild hybridization, new variations on known willow species pop up quicker than scientists can discover or classify them; there are nearly forty known willow subspecies in the Northwest alone. Other common species include Hooker's willow (along the coast), Scouler's willow (in drier zones), and Sitka willow.

LATIN NAME: *Salix lucida* ssp. *lasiandra*

SIZE: Ranges from a multistemmed shrub to a small tree (up to 50 feet)

IDENTIFYING FEATURES: Spear-shaped deciduous leaves that alternate on their branches, shiny on top and soft beneath

GROW REGION: Throughout the Northwest near streams and floodplains

FUN FACT: The forestry and botany communities continue to tweak the name(s) of this willow as understanding of its origins and genetic influences continues to evolve

Fossil Forest

Petrified wood is Washington's state gem, its captivating, colorful whorls of history each a unique work of art. And for those who can't get enough of its quirky beauty, it's time for a trip to the Ginkgo Petrified Forest (Wanapum Recreation Area), a state park along the Columbia River Gorge set aside as a historic preserve in the 1930s after a highway construction crew unearthed some of the rarest specimens ever discovered.

The trees of stone are a good reminder that central Washington used to be a humid, swampy forest millions of

years before, unlike its current arid, sunny climate. Swamp cypress, gingko, maple, walnut, sycamore and horse chestnut all thrived there, along with Douglas fir, hemlock, and spruce at higher elevations.

Mudslides and flooding and other events buried these forests, and chemical reactions eventually transformed wood to stone, the minerals and compounds in the groundwater percolating through the wood to add brilliant patterns. The park now boasts more than fifty species of petrified trees.

Coast Redwood

Along Oregon's rugged and isolated Chetco River watershed in the Rogue River-Siskiyou National Forest is the northernmost grove of colossal coast redwoods in North America, quietly whiling the centuries away amid Douglas fir and Oregon myrtle (see pages 96 and 123) doing the same.

The Redwoods Nature Trailhead is a 1-mile loop through surreal scenery that offers you a chance to pause time and imagine the era when these groves were commonplace throughout the northern hemisphere.

LATIN NAME: *Sequoia sempervirens*

SIZE: They're among the tallest living things on the planet, typically over 200 feet and often taller than 300 feet, often as wide as a truck

IDENTIFYING FEATURES: Evergreen needles with small woody cones only about 1 inch long

GROW REGION: These days, in a narrow band from southwestern Oregon to Monterey, California

FUN FACT: Redwood forests provide habitat for the spotted owl, black-tailed deer, and pileated woodpecker

Pacific Yew

Conifers in the understory often receive less adulation than their towering, light-hogging neighbors. In fact, you may have to play detective to spot a Pacific yew at all while navigating the woods, especially if its trademark self-peeling bark is obscured by moss.

But that moose and those deer watching warily from a safe distance are thankful for this yew's branches, which fill their bellies during lean times. So are the songbirds, who, immune to the toxins of its pretty red fruit, spread its seed far and wide.

LATIN NAME: *Taxus brevifolia*

SIZE: This slow-growing, ancient evergreen eventually reaches 20–40 feet at maturity—after 250–350 years

IDENTIFYING FEATURES: Glossy evergreen needles 1 inch long; its berry-like red fruit has a single large seed

GROW REGION: From southeastern Alaska into California, up to 8,000 feet

FUN FACT: The wood is exceptionally strong and has been used throughout history for tools, canoe paddles, harpoons, and weapons, particularly archery bows

The Tree That Fought (and Beat) Cancer

The Taxus family is infamous for its toxicity. History abounds with tales of woe surrounding yew trees, from accidental livestock and pet deaths to yew compounds being used as a method of suicide or in chemical weapons. But Pacific Northwest native *Taxus brevifolia* (see page 113) found fame for its life-giving properties as well: in the 1960s, researchers from the National Cancer Institute determined an extract from its bark called a taxane had beneficial properties as an anticancer drug.

The lifesaving compound paclitaxel (known commercially as Taxol) blocks cancer cell growth by stopping cell division, resulting in cell death. It's now used to treat ovarian and breast cancer, as well as non-small cell lung cancer, pancreatic cancer, and AIDS-related Kaposi sarcoma.

But the impact of all that bark collection from slow-growing yews led to the species nearly disappearing from the landscape. These days taxanes are largely harvested from cultivated yew trees, a practice that is slowly allowing native yew populations to recover.

Western Red Cedar

This cedar's feathery, fragrant branches are a welcome, frequent sight: you'll find this dominant species looming moodily along shorelines and foothills, sheltering generations of kids from drizzle while waiting for the school bus.

The long-lasting wood provided endless value to Pacific Northwest tribes, who crafted clothing, mats, baskets, implements, and sturdy dugout canoes from the trees and bark. You'll also find it in the Conibear Shellhouse at the University of Washington, where famed boat builder George Pocock looked to those traditional canoes for inspiration to build the speedy Husky Clipper racing shell that helped the Boys in the Boat triumph in rowing at the 1936 Summer Olympics in Nazi Germany.

LATIN NAME: *Thuja plicata*

SIZE: Up to 200 feet tall and as wide as a garage door

IDENTIFYING FEATURES: Flat sprays of needles with relatively small, egg-shaped cones

GROW REGION: British Columbia to Northern California

FUN FACT: The eldest were already growing when George Vancouver first sailed through the region searching for the fabled Northwest Passage in the late 1700s

Western Hemlock

When scanning the tree line, look for the trademark soft, floppy top to spot a western hemlock. Though it may take some time for its crown to reach the top at all: western hemlocks can grow for decades in the shade of common neighbors Douglas fir, Sitka spruce, and western red cedar (see pages 96, 67, and 116), patiently awaiting a windstorm to knock over some taller trees for their turn in the sun.

Once they get that sun, watch out: western hemlocks are a climax species, meaning, if a landscape is left alone, they'll eventually out-shade any other saplings beneath, leaving only fellow shade-tolerant species in the understory.

LATIN NAME: *Tsuga heterophylla*

SIZE: From 100–150 feet tall and 2–4 feet wide

IDENTIFYING FEATURES: Delicate needles of varying size that grow every which way on branches, creating soft, drapey boughs; cones are dainty and plentiful and grow at the ends of branches

GROW REGION: Throughout the moist, temperate forests of Western Washington and northwest Oregon

FUN FACT: *Tsuga heterophylla* is Washington's state tree

Mountain Hemlock

Picture any downhill ski run in the Pacific Northwest and chances are, these slim, fidgety trees are peppered among firs alongside, their flexible branches and trademark droopy crown stooped low with snow like a shepherd's crook until spring.

While cousin western hemlock (see page 119) thrives in the lowlands, mountain hemlock prefers the view from above, growing slowly through the long winters of its life. The eldest top five hundred years; you'll see many of them thriving around the caldera that is Oregon's Crater Lake and the biggest by volume in the Quinault Rain Forest of Olympic National Park.

LATIN NAME: *Tsuga mertensiana*

SIZE: Around 30–100 feet tall, though high winds and extreme cold in subalpine terrain can limit trees to 10 feet or a permanent windswept lean

IDENTIFYING FEATURES: Needles are deep blue green and of equal length; cones are petite and oblong, and often purple when young

GROW REGION: Southeastern Alaska to California, typically above 4,000 feet

FUN FACT: *Mertensiana* comes from Karl Heinrich Mertens, a German botanist who encountered these hemlocks in 1827 while exploring Russian settlements in North America

Oregon Myrtle

It was 1933, the midst of the Great Depression, when the lone bank in the timber and shipbuilding coastal Oregon city of North Bend closed its doors. As in other small towns across America, North Bend leaders crafted their own currency to manage the cash shortage and pay employees out of local treasure: myrtlewood.

The sturdy coins remain legal tender to this day and are prized by collectors for their distinctive grain, which draws its color variations of blond to brown to red from nutrients in its soil.

LATIN NAME: *Umbellularia californica*

SIZE: Tree-sized (40–80 feet tall) in moist environments; shrub-sized under adverse conditions

IDENTIFYING FEATURES: Glossy evergreen leaves with a pungent, bay leaf scent; spring brings small, flat clusters of greenish-yellow flowers

GROW REGION: From coastal southern Oregon to San Diego

FUN FACT: Oregon's champion old-growth myrtle continues to expand along the Myrtle Tree Trail off the Rogue River

AFTERWORD

There are as many "typical" dimensions of trees as there
are trees, it seems. And perhaps that's part of what makes
these species native to the Pacific Northwest so remark-
able: they may generally stick to a pattern, but don't
hesitate to take advantage of optimal soil, light, and other
conditions to flourish where they originally took root.

Whenever possible I used the USDA or US Forest
Service descriptions of typical height and trunk diameter
for consistency. You may notice the word *known* makes
a regular appearance throughout this volume. As with
all aspects of nature and science, our knowledge of what
exists, as well as how and why, is ever evolving. It should
be little surprise when foresters, hikers, photographers, and
others who love the woods chance upon a "new" champion specimen. I take comfort knowing that, like these
trees, perhaps our limits will never fully be discovered.

ADDITIONAL RESOURCES

Books

Flora of the Pacific Northwest: An Illustrated Manual
When you're ready to go all-in on your tree identification habit, pick up this classic by C. Leo Hitchcock and Arthur Cronquist

Northwest Trees: Identifying and Understanding the Region's Native Trees
A less technical classic that will deepen your understanding of the forest, by Stephen F. Arno and Ramona P. Hammerly

Websites and Apps

Burke Herbarium Image Collection (biology.burke.washington.edu /herbarium/imagecollection.php)
An exploration of what you'll find outdoors in Washington State from University of Washington's Burke Museum of Natural History and Culture

HistoryLink.org
Search by tree to learn the role species including Sitka spruce, Douglas fir, and western red cedar played in the region's history, industry, and culture

Native Plants PNW (nativeplantspnw.com)
A deeper look at native flora, including its cultural impact, from long-time horticulturist Dana Kelley Bressette

Oregon State University Department of Horticulture Landscape Plants (landscapeplants.oregonstate.edu)

   Click on Woody Plants of Oregon for step-by-step guides to identifying many native species

Trees Pacific NW

   This photo-rich app by Northwest naturalist Cliff Cantor features extensive descriptions of nearly every common tree you'll encounter on your forays in the woods, making it a snap to identify trees whether up close or from the next ridge over

INDEX

ABOUT THE AUTHOR

Karen Gaudette Brewer was born, raised, and educated in Washington State, where she grew up playing beneath western red cedars, endlessly raking alder leaves, and skiing past forests of Pacific silver fir. She began her career as a journalist with the *Associated Press* and has worked as a writer and editor at the *Seattle Times*, PCC Community Markets, Allrecipes.com, Remedy Health Media, and NerdWallet. In the second grade, she received a poster that cataloged the world's parrots, and it endeared her to taxonomy, a hobby that she is currently imposing on her children.

She makes her home in Seattle with her husband, sons, and cat. This is her second book about the Pacific Northwest.

ABOUT THE ILLUSTRATOR

Emily Poole was born and raised in the mountain town of Jackson Hole, Wyoming. After receiving her BFA in illustration from the Rhode Island School of Design, she returned west to put down roots in the mossy hills of Oregon. She can be found exploring tidepools and cliffsides, gathering inspiration, and making artwork about our fellow species and how to be better neighbors with them.